Education
88

我烦透了！

I Am Bored!

Gunter Pauli

[比] 冈特·鲍利　著

[哥伦] 凯瑟琳娜·巴赫　绘

田　烁　王菁菁　译

上海远东出版社

丛书编委会

主　任：田成川

副主任：何家振　闫世东　林　玉

委　员：李原原　翟致信　靳增江　史国鹏　梁雅丽

　　　　任泽林　陈　卫　薛　梅　王　岢　郑循如

　　　　彭　勇　王梦雨

特别感谢以下热心人士对童书工作的支持：

匡志强　宋小华　解　东　厉　云　李　婧　庞英元

李　阳　刘　丹　冯家宝　熊彩虹　罗淑怡　旷　婉

杨　荣　刘学振　何圣霖　廖清州　谭燕宁　王　征

李　杰　韦小宏　欧　亮　陈强林　陈　果　寿颖慧

罗　佳　傅　俊　白永喆　戴　虹

目录

Contents

一只鸭子正在偷看一座养猪场，他发现里面的猪居然都没尾巴。他转身问路过的山羊："我听说猪长着可爱的、卷卷的尾巴，但是这些猪根本就没有尾巴！这是怎么回事？"

"没有人告诉过你吗？他们一出生，农民就会割掉他们的尾巴。"

A duck is peeping into a pig farm and only sees pigs without tails. He turns to the goat walking past him and asks, "I was told that pigs have funny, curly tails. But these pigs have no tails at all! What happened?"

"Did no one ever tell you? Farmers cut off their tails when they are new-borns."

这些猪根本没有尾巴！

These pigs have no tails at all!

但是，猪不会飞呀！

But pigs can't fly!

"哎哟，那一定会弄疼那些可怜的小家伙们。这就像那些农民过去经常剪掉我们的翅膀，让我们不能飞起来一样。"鸭子回答。"但是，猪不会飞呀！"鸭子一脸迷惑。

"哦，没错，猪不会飞，这是常识，就像地球不是平的。但有些人仍然相信割掉猪的尾巴是有必要的。"

"Ouch, that must hurt those poor little ones. It's like the farmers who used to clip our wings so that we could not fly away," responds the duck. "But pigs can't fly!" he adds, looking puzzled. "Oh no, pigs can't fly, we know that. And the earth is not flat. Still, some people believe it is necessary to cut their tails off."

"我知道人们过去经常会砍掉狗的尾巴，不过这种虐待行为现在已经被法律禁止了。如果禁止人类这样对待狗，那凭什么允许他们这样对待猪呢？"

"噢，如果农民不割掉猪的尾巴，那其他猪也有可能会把它咬下来。那样就更糟了，而且会更疼，还有可能会引发严重的感染。"

"I know that people used to cut off dogs' tails, but at least now that torture is prohibited at last. So, if people are not allowed to do it to dogs, why are they permitted to do it to pigs?"
"Well, if the farmer does not cut off the tail, then it is very likely that another pig will bite it off. That's worse and more painful. It could also cause a bad infection."

可能会引发严重的感染

Could cause a bad infection

永远不会迎来"牙仙子"的造访了

Will never get visited by the Tooth Fairy

"猪会咬掉朋友和同伴的尾巴？我从来不知道猪还这么好斗！"

"还有呢……小猪出生几天后，农民还会拔掉他们的牙齿。"

"这不公平！这样他们就永远不会迎来'牙仙子'的造访了。"

"Pigs biting off their friends' and neighbours' tails? I never knew pigs were so aggressive!"

"There's more … A few days after pigs are born, the farmer also clips their teeth."

"That's not fair! Now they will never get visited by the Tooth Fairy."

"农民担心猪崽会用自己尖利的牙齿伤到他们的妈妈。"

"这可说服不了我！上千年来，猪崽一直都是吸吮妈妈的乳汁长大的。问题在于，人类强迫猪妈妈生了一窝又一窝，还要喂养这么多孩子，这才伤到了她。"

"The farmer is afraid that the piglets will hurt their moms with their sharp teeth."

"That does not convince me! Piglets have been suckling from their moms for millennia. The problem is that these mums are forced to have one litter of babies after the other and feeding them all hurts her."

猪崽会用自己的牙齿伤到他们的妈妈

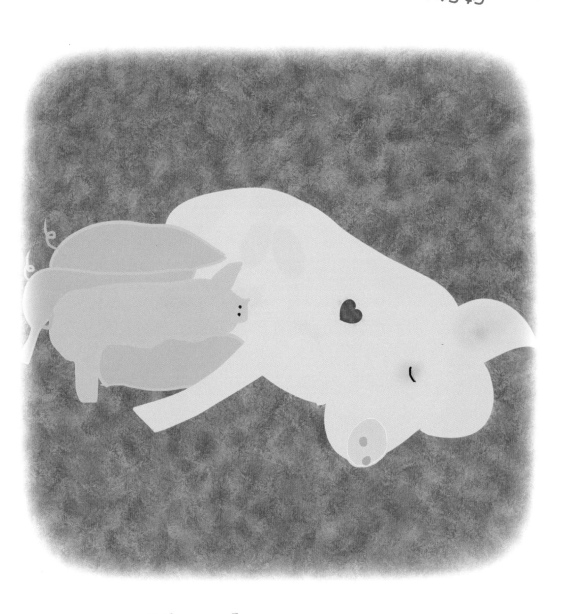

Piglets will hurt their moms with their teeth

母鸡会互相伤害？

Hens hurt each other?

"嗯，你知道这已经不仅仅是饲养这么简单了，这是工业化生产。同样的情况也发生在母鸡身上，她们也会互相伤害。"

"母鸡会互相伤害？公鸡好斗，母鸡也好斗？她们是世界上最友好的动物，我才不相信呢。"

"Well, you know this is just not farming anymore.
This is industrial production.
The same is happening to the hens.
They also hurt each other."
"Hens hurt each other? Cocks fight, but hens?
They are the friendliest animals on Earth. I don't believe this."

"哦，是的，当很多小鸡被圈养在一个狭小的空间里，他们的喙会被剪得平平的，以防他们啄掉其他鸡的毛，甚至把其他鸡杀死。"

"你确定这不是公鸡为了求偶而发起的争斗？"

"Oh yes, when chickens are confined to a small space, their beaks are trimmed to avoid them pulling out each other's feathers or even killing one another."

"Are you sure these aren't males fighting about a female?"

他们的喙被剪得平平的

Their beaks are trimmed

我认为，他们只是变得烦躁了

I think they simply get bored

"不是，这些都是会下蛋的鸡，她们都是母鸡。"

"我从来不知道母鸡也会变得如此暴躁，以至于农民不得不修剪她们的喙。"

"鸡和猪天性并不好斗。我认为，他们只是变得烦躁了。"

"No, these are all egg-laying chickens. They are all female."

"I never knew hens could get so cross with each other that the farmer had to cut off their beaks."

"Chickens and pigs are not aggressive by nature. I think they simply get bored."

"烦躁？他们当然会烦躁。他们整天待在一个小笼子里，吃着自己不喜欢吃的食物，只等着被装进卡车运往远处进行加工。对我而言，这听起来就不是什么愉快的事！我不会烦躁，我只会发疯！仅仅是想象这种悲惨的生活，就足以让我有咬掉你尾巴的冲动！"

……这仅仅是开始！……

"Bored? Of course they are bored. Sitting in a small cage, being fed something they don't like to eat, only to be sent off in a truck to be processed far away. That doesn't sound very pleasant to me! I wouldn't get bored, I would get mad! Just the thought of such a bad life gives me the urge to bite off your tail!"

… AND IT HAS ONLY JUST BEGUN!…

······这仅仅是开始！······

...AND IT HAS ONLY JUST BEGUN!...

你知道吗?

When their environment is uninteresting and unstimulating, pigs start biting each other's tails. As a result, their tails are cut (tail docking) and their teeth are clipped.

当周围环境枯燥乏味时，猪就开始互相咬尾巴。所以，人们干脆割掉它们的尾巴（断尾处理），拔掉他们的牙齿。

The renewable energy lobby's slogan "Pigs can Fly, the Earth is a Square and Nuclear Energy is Safe" became very popular after the Three Mile Island nuclear meltdown (1979) and the Chernobyl disaster (1986).

可再生能源的游说口号——"猪是会飞的，地球是方形的，核能是安全的"——在三里岛核泄漏事故（1979）和切尔诺贝利核电站事故（1986）发生后流行起来。

Pigs are forest and swamp creatures. Since pigs cannot cool themselves, they have difficulty living in the desert. They have habit of rolling around in urine and excrement to keep cool.

猪是生活在森林和沼泽中的物种。由于猪不能自我降温，它们很难在沙漠中生存。猪会在尿液与粪便中打滚来保持凉爽。

Cutting dogs' tails was thought to prevent rabies. This was later extended to ear docking. However, the real reason for cutting some dogs' tails and ears is tradition and aesthetics. Most countries have banned this practice.

人们曾经认为割掉狗的尾巴可以预防狂犬病，后来又发展为割掉耳朵。然而，割掉狗的尾巴、耳朵的真实原因不过是传统习俗和审美的需要。很多国家已经禁止了此类做法。

Piglets are born with eight needle teeth, which are later replaced by permanent teeth. The piglets establish a teat order, and during the first three weeks of their life, they solely depend on mother's milk for nutrition, always nursing on the same teat.

猪崽出生时有八颗尖牙，然后会换成恒牙。猪崽有哺乳秩序，在出生后的三周内，它们只能依赖母乳来获取营养，每个猪崽通常只固定在一个乳头上吃奶。

Chickens will sometimes pluck each other's feathers until blood is drawn. They will do this when it is too hot, too crowded, or if there is not enough fresh air. Since chickens have the tendency to imitate each other, the whole flock may start to peck at one other aggressively.

小鸡有时会相互啄毛直到出血。在感到太热、太挤，或呼吸不到新鲜空气时，小鸡就会有这样的行为。因为鸡有相互模仿的习性，因此，当一只鸡被激发出了斗性，整个鸡群就会开始互啄起来。

When chickens are stressed and pecking at each other, the pecking can be stopped by feeding the hens fresh grass, hay, or lettuce or by dimming the lights in the hen house.

当鸡群感到压力并互相攻击时，给母鸡喂些鲜草、干草或生菜，或者调低鸡舍的光线，上述攻击行为就会停止。

Boredom reminds some people of the perceived meaninglessness of their existence. This feeling has been recognised by psychologists as a reason for aggression.

枯燥乏味会让一些人感觉自己的存在没有意义。心理学家认为这种感觉是引发攻击行为的原因之一。

Would you rather cut off a pig's tail to avoid it being bitten off or find ways to make the pig happy?

你是愿意靠割掉猪的尾巴来避免猪互相撕咬，还是希望找到让他们快乐起来的方法？

Do you consider a dog more beautiful when its tail and ears are cut?

你觉得当狗的尾巴和耳朵被割掉后，狗会变得更好看吗？

Do you like the story of the Tooth Fairy? Do pigs have a Tooth Fairy as well?

你喜欢牙仙子的故事吗？猪也会有牙仙子吗？

When you are bored, are you relaxed or are you anxious?

当你感到无聊时，你是觉得轻松还是焦虑？

How many of your friends know that people cut the tail and ears of some dogs, the tail and teeth of pigs, and the beaks of chickens? Ask around to find out. To your surprise, you may learn that hardly anyone knows anything about this. Collect some pictures of dogs or pigs whose tails were cut off and ask people if they think this is really necessary. What do you think the answer will be?

你的朋友中有多少人知道人类会割掉狗的尾巴和耳朵，割掉猪的尾巴，拔掉猪的牙齿，剪平鸡的嘴巴？问一下周围的人并找到答案。让你吃惊的是，你可能会发现几乎没有人了解这些情况。收集一些被割掉尾巴的狗或猪的照片，问问人们是否真的有必要这么做。你认为答案会是怎样的？

学科知识

Academic Knowledge

生物学	尾巴的功能包括助力起飞，改善平衡，游泳时把握方向，抓握树枝（卷尾）；猪出生时有乳牙，一段时间后会换成恒牙。
化 学	喙上覆盖有一层皮肤，这层皮肤可以产生角蛋白；鸟类喙上的角蛋白会变干、凝结，让喙部更加坚硬耐用。
物 理	牙齿和喙的角度已进化为可以满足特定需求的形态，尤其是对于鸟类而言，它们要适应不断变化的饲养方式和饲料的变化；一些动物皮肤上的斑点是一种物理效应，从来没有两只动物有相同的斑点。
工程学	动物尾巴、牙齿和喙的几何形状为工程师设计飞机、锯和高铁提供了灵感。
经济学	从农业向农业产业的演进。
伦理学	双重标准的概念：为什么禁止以某种处理方式对待狗，却允许以这种方式对待猪。
历 史	达·芬奇曾预言，在未来，杀害动物将被视为和谋杀人类一样恶劣；狄更斯在他的小说《荒凉山庄》（1852年）中写下了"无聊至死"（"bored to death"）的名言。
地 理	文化不敏感与虐待动物：某些国家以特定方式对待动物的传统看起来是一种虐待，比如，西班牙斗牛，还有在比利时喝啤酒时在酒杯里放一条活鱼。
数 学	利用非线性矢量函数，图灵第一个论述了动物皮肤上的图案是如何形成的。
生活方式	以某种特定方式对待动物使它们看起来更美（从人类的视角）、更健康、更安全，却无视动物因此遭受的痛苦；无聊有时会促使人们吸烟或酗酒。
社会学	剧作家易卜生在《海达·高布乐》中展现了持续且强烈的无聊感容易变成攻击行为的现象；由于媒体或利益集团的宣传，人们会相信那些不正确的东西；动物福利倡导者接受为实现人类目的而利用动物的观点，但希望减少动物的痛苦；动物权益保护人士要保护动物，结束其作为财产的地位，确保它们不再被用作商品。
心理学	无聊会激发一些人天生的冲动思维，这些人往往是寻求新鲜经历却未能如愿的人；无聊，当然还有其他因素，会降低预期寿命；无聊也会激发好奇心和更多的联想以及创造性思维方式。
系统论	系统科学在思维中的运用正如它在物流和生态系统中的应用一样：烦躁会导致古怪的行为。

情感智慧
Emotional Intelligence

山羊

山羊冷静而关切，希望分享自己的知识和想法。他准备开一个玩笑，但遵守规则的限制。他表现出了实用主义，解释了为什么尾巴必须被割掉（以免带来更多疼痛）。山羊对这些知识进行了拓展，指出猪崽不仅没了尾巴，而且还没有了牙齿。山羊接受了人类制造这种痛苦背后的逻辑。他们的对话继续进行，进而揭开了更多的痛苦——鸡因喙被剪平而承受的痛苦。然而，根据山羊的观点，这是不可避免的，因为要确保鸡群不互相伤害。最后，山羊承认，不管是鸡还是猪，他们都不具有攻击性，他们的不良行为是由工业化居住条件所带来的无聊感而导致的。

鸭子

鸭子的求知欲很强，能思考现象背后的问题。他表现出情感共鸣，从猪和狗的境遇联想到自身处境。由于对待狗的禁律却可以施用于猪，鸭子质疑人类的双重道德标准。当山羊解释了拔牙的事情后，鸭子担心猪崽们永远不会迎来"牙仙子"的造访。鸭子并没有被山羊的解释说服，他认为人类没有理由去虐待这些可怜的动物。针对山羊的解释，鸭子的反对态度变得越来越明确，他怀疑一切，也表现出了自己的无知。当山羊最后承认问题根源不在于动物的不良行为，而是人类的错误时，鸭子表达了自己的愤怒。

艺术
The Arts

《三只小猪》是一篇著名的童话，讲述了三只猪和它们的房子的故事。让我们给这个故事换一副新面貌，画三座不同风格的房子，然后在房子前面画三只小猪，着重突出它们的牙齿和尾巴。

思维拓展
Systems: Making the Connections

烦躁的人面临着选择：一种方式是通过停止重复性和无意义的活动，为生活注入刺激来变得有创造力，并改善生活。另一种选择是从无趣到攻击，使用武力和打斗来寻找兴奋点，摆脱空虚。动物也有同样的心态。许多动物被驯化了，经过五千至上万年的进化，它们变成人类忠实而友好的伙伴。现代学者持有这样一种观点——人类是地球的守护者，必须要关爱地球和大自然。显而易见的是，我们增加动植物产出的努力已经达到了这样一种程度——有些人认为它是非常高产的，有些人则认为这是一种虐待。瑞士早在1992年就以立法形式禁止了对鸡喙的修剪处理，而很多国家现在才刚刚开始实施法律政策来规范工业化农场养殖。温顺动物的生存条件往往比奴隶的生存条件还要恶劣，这些动物对此的反应让我们想起了达·芬奇在16世纪的预言——人类对待动物的方式不久将会施于人类自身。动物因囚禁而变得烦躁，继而产生攻击行为的过程是反自然的，并产生了一系列连锁反应。在这些连锁反应中，人类浑然不觉（希望是这样）给动物们带来了更多的痛苦和绝望。食物的生产需要以一种和谐的方式进行。这不仅仅是要确保高品质的生活，也是为了保证随着地球人口的增加以及更多的人成为中产阶级，他们会改变自己的饮食习惯，创造一种可持续发展的供应链，不突破地球承载力的极限。

动手能力
Capacity to Implement

当我们身边充斥着消极情绪时，保持积极向上几乎是不可能的。虐待动物的行为几乎遍布全世界，这甚至是一些人成为素食主义者的一个原因。问问周围的人，在家庭成员和朋友中有没有素食主义者。然后探究一下是否有人是因为不能忍受食用遭受虐待的动物的肉而变成素食主义者的。

故事灵感来自
This Fable Is Inspired by

卡尔·路德维希·施魏斯福尔特
Karl Ludwig Schweisfurth

　　第二次世界大战结束后不久，卡尔·路德维希·施魏斯福尔特在美国的一些大型屠宰场接受了屠宰训练。这段经历帮助他将自己在德国赫塔的家族生意发展为欧洲最大的肉食加工企业。他的公司一周可以屠宰 5 000 头牛、25 000 头猪。他认为，那些遭受过虐待的动物肉的营养价值和肉质从来都不怎么好。他的两个儿子对于接管这个生意没有兴趣，他们认为这个工作是不人道的，因此他决定把这份生意卖给雀巢公司。他再度创业，将动物饲养工作和屠宰场（动物们在这个屠宰场以一种比较有尊严的方式被屠宰）、烘焙坊、奶酪加工厂、酿酒厂、饭店和旅馆等结合在一起，以最高的生态标准来运营。卡尔·路德维希创建了一个以其姓氏命名的基金会并组建了一个团队去学习如何生产高品质且环境友好型食品。他们的目标是所售产品的价格要为大多数人所接受。

图书在版编目（CIP）数据

冈特生态童书.第三辑修订版:全36册:汉英对照 /
(比)冈特·鲍利著;(哥伦)凯瑟琳娜·巴赫绘;
何家振等译.—上海:上海远东出版社,2022
书名原文:Gunter's Fables
ISBN 978-7-5476-1850-9

Ⅰ.①冈… Ⅱ.①冈… ②凯… ③何… Ⅲ.①生态环
境–环境保护–儿童读物—汉、英 Ⅳ.①X171.1-49

中国版本图书馆CIP数据核字(2022)第163904号
著作权合同登记号图字09-2022-0637号

策　划　张　蓉
责任编辑　程云琦
封面设计　魏　来　李　廉

冈特生态童书
我烦透了！

[比]冈特·鲍利　著
[哥伦]凯瑟琳娜·巴赫　绘
田　烁　王菁菁　译

记得要和身边的小朋友分享环保知识哦！
八喜冰淇淋祝你成为环保小使者！